ALBEMARLE COUNTY [VIRGINIA] ROAD ORDERS

1744-1748

Virginia Genealogical Society
Richmond, Virginia

Published With Permission from the

Virginia Transportation Research Council
(A Cooperative Organization Sponsored Jointly by the Virginia
Department of Transportation and
the University of Virginia)

HERITAGE BOOKS
2008

HERITAGE BOOKS
AN IMPRINT OF HERITAGE BOOKS, INC.

Books, CDs, and more—Worldwide

For our listing of thousands of titles see our website
at
www.HeritageBooks.com

Published 2008 by
HERITAGE BOOKS, INC.
Publishing Division
100 Railroad Avenue #104
Westminster, Maryland 21157

International Standard Book Number: 978-0-7884-3664-2

PREFACE

The Virginia Highway and Transportation Research Council is a cooperative organisation sponsored jointly by the Virginia Department of Highways and Transportation and the University of Virginia and is located on the Grounds of the University at Charlottesville. The Council engages in a comprehensive program of research in the field of transportation. As a part of its program the Council, in December 1972, began research on the history of road and bridge building technology in Virginia. The initial effort was concerned with truss bridges; a complementary effort concentrating on roads got underway in October 1973.

The evolution of the road system of Virginia is in many ways inseparable from the social, political and technological developments that form the history of the Commonwealth. Despite this, there are few extant serious works on the history of roads in Virginia. Those which have been produced focus on internal improvements and turnpike development before the War Between the States. Little has been done on the period from Reconstruction through the creation of the system of state highways in the earlier part of this century.

Accordingly, it was decided to investigate the development of the roads of Albemarle County during the period 1725-1925 as a pilot project, and to use this experience to produce a "History of Albemarle County Roads" and a procedural handbook for the writing of Road Histories. During the early stages of this project it was necessary to examine and extract all the Road Orders for the Counties from which Albemarle was formed as well as the Orders for Albemarle where it still contained the Counties of Amherst, Buckingham, Fluvanna, Nelson, and parts of Appomattox, Bedford and Campbell. The Road Orders concerning Albemarle will ultimately be published with the Road History but the broad applicability of those for Goochland, Louisa and early Albemarle, and the opinions of various authorities throughout the state who have examined them, indicate that they should have separate publication in order to make them generally available to individual scholars through libraries and educational institutions. Therefore, this publication has been prepared.

ALBEMARLE COUNTY ROAD ORDERS 1744-1748

By

Nathaniel Mason Pawlett
Faculty Research Historian

INTRODUCTION

The roads are under the government of the county courts, subject to be controuled by the general court. They order new roads to be opened whenever they think them necessary. The inhabitants of the county are by them laid off into precincts, to each of which they allot a convenient portion of the public roads to be kept in repair. Such bridges as may be built without the assistance of artificers, they are to build. If the stream be such as to require a bridge of regular workmanship, the court employs workmen to build it, at the expence of the whole county. If it be too great for the county, application is made to the general assembly, who authorize individuals to build it, and to take a fixed toll from all passengers, or give sanction to such other proposition as to them appears reasonable.

Thomas Jefferson, *Notes on the State of Virginia*, 1781.

The establishment and maintenance of public roads was an important function of the County Court during the colonial period in Virginia. Each road was opened and maintained by an Overseer or Surveyor of the Roads charged with this responsibility and appointed by the Gentlemen Justices. He was usually assigned all the "Labouring Male Titheables" living on or near the road for this purpose.

Major projects, such as bridges over rivers, demanding considerable expenditures were executed by Commissioners appointed by the Court to select the site and contract with workmen for the construction. Where bridges connected two counties, a commission was appointed by each and they cooperated in executing the work.

The Road Orders contained in the Albemarle County Court Order Book covering the period 1744-1748 are the principal evidence concerning the early roads remaining in the records of the County. During this period, Albemarle stretched southward to Appomattox River and west to the "Ledge" or Blue Ridge and contained all or part of the present Counties of Albemarle, Amherst, Appomattox, Bedford, Buckingham, Campbell, Fluvanna and Nelson. Insofar as possible, all the Orders were extracted verbatim and the capitalisation, spelling and punctuation have been reproduced without any attempt at correction or consistency.

THE DEVELOPMENT OF ALBEMARLE COUNTY

Note: As originally published in paper format, this volume included maps showing the evolution of the county. Maps are not included in the revised/electronic version due to legibility and file size considerations. Instead, a verbal description is provided.

By the 1720s, the area that is now Albemarle County was part of the western reaches of Goochland County (created in 1728 from Henrico County) and Hanover County (created in 1721 from New Kent County). In 1742, the western section of Hanover was cut off as Louisa County (then including the northern third of modern-day Albemarle County).

Albemarle County was created in 1744 from Goochland County. In its original form, Albemarle contained the southern two-thirds of modern Albemarle County, the entirely of the modern counties of Amherst, Buckingham, Nelson, and Fluvanna, and parts of Appomattox, Campbell, and Bedford counties. Albemarle's boundaries were considerably reduced with the creation of Buckingham County (then also containing part of modern Appomattox County) and Amherst County (then also containing present-day Nelson County) in 1761. The year 1762 brought a slight northward expansion of the county boundaries, with the addition to Albemarle of the western portion of Louisa County. This brought Albemarle's western, southern, and northern boundaries to their current locations. With the last reduction in its territory, the creation of Fluvanna County in 1777, Albemarle reached its present size.

28 March 1745 O.S., p. 4
Road./. [Italicized entries are marginal notes]
Ordered that John Henderson be Summoned to Appear at next Court to Shew cause if any he hath why a road (destroyed) Cleared through the Land of the said Henderson, from the Three Notched Road and so to Hardway River./.

28 March 1745 O.S., p. 5
Road./.
James Defoe is Appointed Surveyor of the highway from Number 12: to Number 18. and the Male Tithables that formerly belonged to the said Road are Ordered to Assist the said Defoe in clearing the same

28 March 1745 O.S., p. 5
Road./.
John Woody is Appointed Surveyor of the High way from Number 18 to the County Line on the three Notched Road. and the male Tithables that formerly belonged to the said Road are ordered to Assist the said said Woody in clearing the same

28 March 1745 O.S., p. 5
Road./.
Charles Lynch Gent: is Appointed Surveyor of the High Way from the late Secretary's foard to Number 12., and Likewise of the road to the said Lynches ferry: and the same male Tithables that formerly worked under the said Lynch, are ordered to Clear both roads./.

28 March 1745 O.S., p. 8
Roads./.
John M^c.Cord is Appointed Surveyor of the High Way from D.S. to W.M. and the Male Tithables that formerly belonged to the said Road./. together with Thomas Hughs David Addams Daniel Lattimore and John Cambhill are Ordered to Assist the said M^c.Cord in Clearing the same./.

28 March 1745 O.S., p. 9
Roads./.
William Morrison is Appointed Surveyor of the Highway from the uper end of M^c.Cords Road to Thomas Morrisons, and the Male Tithables that formerly belonged to the said Road, are Ordered. to Assist the said Morrison in Clearing the same./.

25 April 1745 O.S., p. 10
Pet°. for Road./.
On the Petition of Benjamin Wheeler and others, for Road (from the said Wheeler's into the four Chopt Road; to Wood's Gapp) leave is granted them to clear a road According to the said Petition./.

25 April 1745 O.S., p. 10
Road./.
James Taylor is Appointed Surveyor of the High Way, in the Room of John Thrasher, and the same hands that were Appointed to work under the said Thrasher. are Ordered to Assist The said Taylor In Clearing the same./.

25 April 1745 O.S., p. 10
Pet^o. for Road./.
On The Petition of David Lewis for a road over Capt. Charles Lynches foard, leave is granted him to Clear a road Accordingly./.

25 April 1745 O.S., p. 10
River View'd & c.
Ordered that the settlement of the Court House that was refered to the Consideration of this Court. be Continued til the next and that Peter Jefferson Allen Howard Charles Lynch Thomas Bellew & William Cabell Gent: in the meantime do View the River, and make their report thereof to the next Court./.

23 May 1745 O.S., p. 13
Road./.
On the motion of Robert Rose Clerk, leave is granted him to Clear a Road from his Plantations on Tye River to Leak's Plantation

23 May 1745 O.S., p. 13
Road./.
On the motion of John Harvie leave is granted him to Clear a road from the Mouth of Tye River to the Branches of Harris's Creek./.

27 June 1745 O.S., p. 23
Road's
Ordered that the Surveyors formerly Appointed for the several high ways within this County be Continued./.

27 June 1745 O.S., p. 23
Road./.
Andrew Wallace is Appointed Surveyor of the High Way from D. S. to Mitchams River. and Archibald Woods, Jeremiah Marris, William Shaw, Robert Mannely, John Dickey William Wallace Mirlock M^c.dowell Micah Woods jun^r. Micah M^c.Dowell Anthony Osbrook John Lawson John Cowan William Little and Robert Anderson Are Ordered to Assist the said Wallace in clearing the same./.

27 June 1745 O.S., p. 23
Road./.
William Cabell. Gent. is Appointed Surveyor of the High Way from the ffoard of Tye River to the Court. House, the Male Tithables betwixt the said Road and the River; and allso. the Male Tithables

Convenient to the said Road belonging to the Rev^d: M^r. Stith M^r. Jn°. Harris, Charles. Lavinder, John Isham, and Samuel Burks./. Are Ordered to Assist the said Cabell, in Clearing the same./.

27 June 1745 O.S., p. 23
Road./.
John Anthony is Appointed Surveyor of the High Way, (to be laid off be Charles Lynch Gent). from the Court. House to Martin Kings Road, all the Male Tithables, from the mouth of Hardwar to the Court House, (that are not allready Appointed) Are Ordered to Assist the said Anthony in Clearing the same

27 June 1745 O.S., p. 24
Road./.
Ordered that A Road. be Cleared from the place Martin Kings Road come into the Three Notched Road. by Ellis Hues'/ to the County Line. towards Louisa Court House, that the Surveyor, and Hands, belonging to Martin Kings Road, Do Clear the same./.

And that the Male Tithables of Capt. Thomas Meriwether Jn°: Edmonds and Samuel Crawley. be Added to Jn°. Anthony's Gang & c.

27 June 1745 O.S., p. 24
Road.
Thomas Bellew Gent; is Appointed Surveyor of the. High Way to be laid off by the said Bellew. from the Court House to the Road formerly Cleared by Allen Howard Gent: and from thence to be Carried by the said Bellew. the best way to Slate River And all the Male Tithables betwixt the said Road and the Fluvanna River are Ordered to Assist the said Bellew in Clearing the same

27 June 1745 O.S., p. 24
Road./.
John Cannon is Appointed Surveyor of the High Way to be laid off by the said Cannon from the place Capt. Bellew's Road comes into Slate River the nearest and best way to Glovers Road. And all the Male Tithables betwixt the said Road and. the Fluvanna River are Ordered to Assist the said Cannon in Clearing the same./.

27 June 1745 O.S., p. 24
Road./.
Allen Howard Gent. is Appointed Surveyor of the High Way on the South side of River from the Court House up the Country, to the Road formerly Cleared by the said Howard and from thence to SycoffLer Island Creek. and all the male Tithables betwixt the said Road and the fluvanna River are Ordered to Assist the said Howard in Clearing the same./.

27 June 1745 O.S., p. 24
Road./.
Ordered that Thomas Bellew. Gent. and Richard Tayler. do Mark Out a Road. from the Court. House, to Meriday Mannings foard on Slate River. And that William Allen Gent. do Carry on the said Road. the nearest. and best way to Horn Quarter./.

27 June 1745 O.S., p. 25
Roads./.
Thomas Joplin is Appointed Surveyor of the High Way from Rock fish to Hardwar in the Room of John Johns. and the Hand formerly Appointed to work on the said Road. Except. Capt. Nevell's. are Ordered to Assist the said Joplin in Clearing the same./.

27 June 1745 O.S., p. 25
Road./.
John Key is Appointed Surveyor of the High Way, from the End of A Road which is Cleared, to John Keys Mill, the nearest. and best way to the Late Secretarys ffoard. and the Male Tithables of William Womac, Daniel Holladay, Robert Daulton, William Hind, John Key, and, Larkin Smith are Ordered to Assist the said Key in Clearing the same./.

27 June 1745 O.S., p. 25
Road./.
Thomas Jones is Appointed Surveyor from the fork of Road to Rock fish foard and allso of the Road to Swans Creek. the Gang of Hands formerly Appointed on the said Road as well as the present gang are Ordered to Assist the said Jones in Clearing the same./.

27 June 1745 O.S., p. 25
Road./.
John Fenley is Appointed Surveyor of the High Way. James Fenly was formerly Appointed Surveyor of, and the same hands that were Ordered to Work under James. are now Ordered to Assist John Fenly in Clearing the said Road./.

27 June 1745 O.S., p. 25
Diret^n. & c.
The Surveyors of the Several high Ways within this County Are Ordered to set up Posts with Direction at the several forks of Roads with in their respective Districts; According to Law, and that the said Directions be set up at Least ten feet from the Ground./.

25 July 1745 O.S., p. 33
Scotts. ferry.
On the motion of Daniel Scott leave is given him to keep a ferry from the Court House Landing to the Opposite side he giving Cond. and Security as the Law requires./.

25 July 1745 O.S., p. 33
Scotts Ord[y]. Licence)
On the motion of Daniel Scott leave is given him to keep an Ordinary at the Court House giving Bond and Security as the Law Requires./.

Daniel Scott together with Edmond Gray enter into and Acknowledge Bond in the Sum of ten thousand pounds of Tobacco on Cond: that the said Daniel Scott shall Constantly provid in his said Ordinary good wholesome and cleanly Lodging and Diet for Travellers. and Stablage & c. for Horses. During the Tearm of One Year & c.

25 July 1745 O.S., p. 34
Road./.
On the Petition of the Inhabitants of the Uper part of Mitchams River leave is granted them to Clear a Road from Rock fish Gapp the nearest and best way to D.S. road./.

25 July 1745 O.S., p. 34
Road
Ordered that the Road from Robert Davis's ffoard to Howard Cashe's be Cleared and that the Inhabitants above the Red Mountains do Clear the same./.

25 July 1745 O.S., p. 34
Road./.
Ordered that a Road be Cleared from the ffoard on the Mitchams River to Michael Woods Gapp. on the Blue Mountains. and that the Inhabitants above Mitchams River that are not allready Appointed do Clear the same./.

25 July 1745 O.S., p. 34
Road./.
On the Petition of the Inhabtitants in the ffork of James River Leave is granted them to Clear a Road from the Court House to Shepards ffoard from thence to Benjamin Woodsons ffoard. and from thence to Doct[r]. Hopkins's Road./.

Ordered that Lazarus Dameron George Hilton and Benj[n]. Woodson be Overseers of the Said Road./.

22 Aug. 1745 O.S., p. 45
Lynch's ferry Bond & c.
On the motion of Charles Lynch Gent; leave is given him to keep a Ferry from his Land a Cross the North River to the Opposite side the said Lynch giving Bond and Security as the Law requires./. Charles Lynch together with William Cabbell Gent. enter into and Acknowledge Bond & c. on Condition the said Lynch shall keep a ferry as the Law requires

22 Aug. 1745 O.S., p. 45
Road./.
On the Petition of Joseph Kinkead and Andrew Wood leave is given them to Open a Road from the said Kinkeads House to Davis Stockton Mill according to the said Petition./.

23 Aug. 1745 O.S., p. 60
Road./.
William Woods is Appointed Surveyor of the Road from the Foard on Mitchams River to Michael Woods's Gapp on the Blue Mountains. and the Inhabitants above Mitchams River not allready Appointed on Other Roads are Ordered to Assist the said Woods in Clearing the Same

23 Aug. 1745 O.S., p. 60
Road./.
On the motion of Charles Lynch Gent: Mathew Graves is Appointed Surveyor in the room of the said Lynch of the Road from the Secretary's Foard to twelve Mile tree And the Gang fformerly Appointed to Work under the said Lynch are Order'd to Assist the said Graves in Clearing the same.

26 Sept. 1745 O.S., p. 61
Road./.
Ordered that William Allen be Surveyor of the Road from Meriday Mannings to Horn Quarter and that the Hands Convenient to the said Road do Assist the said Allen in Clearing the same./.

26 Sept. 1745 O.S., p. 64
Road./.
Ordered that John Graves be Overseer of the road from the mouth of Tye River to the Branches of Harris's Creek and that all the Male Tithables between Buffiloe River and the Fluvanna (Charles Lavinder only excepted) do Assist the said Graves in Clearing the Same./.

26 Sept. 1745 O.S., p. 64
Road./.
Ordered./. that George Hilton be Overseer of the road from the Court House to Shepards foard from thence to Benjamin Woodsons ffoard and from thence to Doct^r. Hopkins's Road./. and that all the male tithables from the Mouth of Hardwar River to the Point of fork, thence up the North River including Mr. Thomsons Tiths however and all the inhabitants from thence up Cunningham Creek and the branches there of from thence to Hardwar River and from thence to Benjamin Woodsons.

26 September 1745 O.S., p. 65
Road./.
On the Motion of Joseph Thomson Gent: leave is given him to Clear a Road from his Old House to Doct^r. Hopkin's Road./.

26 Sept. 1745 O.S., p. 65
Road./.
Ordered that Moses Higginbottom be Overseer of a Road from the said Higginbottom's Mill a Cross Buffiloe to M^r. Harveys Road and that the Male Tithables Between Buffiloe and the Secretarys Mountains do Clear the same and that the said Gang do Join Jno Graves's to the mouth of Tye River./.

26 Sept. 1745 O.S., p. 65
Road./.
Ordered that John Key do Continue his road from Keys Mill to the Secretarys ffoard and that the Male Tithables of M^r. Meriwether Col^o. W^m. Meriwether and Jno Kerrs be added to the hands formerly Appointed under the said Key do Clear the same./.

27 Sept. 1745 O.S., p. 78
Road./.
Ordered that John Anthony with the Gang formerly Appointed to work under him do Continue to keep the Road he was formerly Appointed, to Clear and repair.

28 Nov. 1745 O.S., p. 81
Road.
On the Motion of William Harris. leave is given him to Clear a Road from his Plantation on Green Creek to the South River on the lower side of Barrengers Creek

29 Nov. 1745 O.S., p. 85
Road.
Ordered the Road from D. S. Road to Capt. Charles Lynch's ferry be a Publick Road and that the Gang formerly Appointed to Work under Capt. Lewis do Clear the same./.

29 Nov. 1745 O.S., p. 85
Road.
Ordered that John Anthony and the Hands formerly Appointed to work under him and Lazarus Dameron with his hands do Assist in Clearing the Road from the Court House to Shepards foard on Hardwar River./.

23 Jan. 1745 O.S., p. 88
Road./. View^d.
On the Petition of Thomas Walker for a Road from the north Garden through Ivey Creek Pass to Rock fish Road. Order'd that William Sudworth Hugh Dobbins and John Lyon or Any two of them do View the same and make their Report there of to Next Court./.

23 Jan. 1745 O.S., p. 92
Road.
Ordered that Henry Burras. Overseer George Tayler and John Heard or any two of them do View the Road from the 3 notch'd road to Rooks's Foard, from thence the best way a Cross Buck Island Creek to Martin King's and John Anthony's Roads {according to the Petition of Tho[s]. Bibb & Tho[s]. Walker and that they report their Oppinion thereof to Next Court./.

27 March 1746 O.S., p. 102
Road. & c
Ordered the Road from Samuel Glovers to Horn Quarter be Publick Road, and that all the Inhabitants on the South side Slate River. that are convenient to the same do Assist in Clearing it and that Samuel Glover be Overseer.

27 March 1746 O.S., p. 102
Road & c
Ordered that the Road from Three Notched Road to Rooks's Foard that was formerly Viewed by Order of this Court be a Publick Road, and that the hands belonging to the late Secretary below the Mountains do Clear the same Thomas Bibb Overseer.

27 March 1746 O.S., p. 102
Road & c
Ordered the Road from Glovers Quarter to Slate River be Continued as the Old Road went, that John Cannon be Overseer. And that all the male Tiths within said Road do Assist in Clearing the same./.

27 March 1746 O.S., p. 102
Road & c
Ordered that Benjamin Woodson's Majr John Henry's and the uper part of Miles Cary's Hands and all others who live Contiguous between Samuel Burks and the Road do Clear According to a former Order./.

27 March 1746 O.S., p. 102
Road.
Ordered Henry Martin do Clear the Road from the Place it comes into Hopkins's road to the County line and that his Old D[r]. Hopkin's Col[o]: Martin's and M[r]. Adams's Hands do Clear the Same./.

27 March 1746 O.S., p. 102
Road & c
Ordered Capt James Martin be Overseer of One half of the Road Willia Morrison was formerly Overseer of, and that one Half of the Hands living on the said Road do Assist the said Martin in Clearing the same./.

27 March 1746 O.S., p. 103

Road. & c

Ordered Joseph Smith be Overseer of the Horn Quarter Road from the County Line to Glovers Road and that M[r]. Cary's hands at Hatchers Creek. Anthony Dinnies. John Hunters. and Robert Kings do Clear the same.

27 March 1746 O.S., p. 103

Road & c.

Ordered that all the Hands from the South Side Slate River that are not all ready Ordered under other Surveyors. do work under William Allen on the Road he is Surveyor on./.

27 March 1746 O.S., p. 103

Road & c.

Ordered Robert Adams be Overseer of the Road Thomas Martin was formerly Overseer of and that the Hands formerly Ordered to Work under the said Martin do now work under the said Adams.

27 March 1746 O.S., p. 103

Road & c

Ordered William Barnett be Overseer in the room of Thomas Herbert and that he have the same hands.

27 March 1746 O.S., p. 103

Road & c

Ordered the Road from the Court House to Shepards foard be Carried round the Ridge. John Lewis be Overseer of the same in the room of Lazarus Dameron & that he have the same Hands./.

27 March 1746 O.S., p. 103

Road & c

Ordered Andrew M[c]:Williams be Overseer in the Room of William Woods and that he have the same Hands./.

12 June 1746 O.S., p. 121

Road

Ordered that Samuel Glover be Overseer of the Roads his father was formerly Overseer of and that the same hands do now work under him who were formerly appointed to work under his father.

12 June 1746 O.S., p. 122

Road.

Ordered the Petition of Charles Anderson and Others for Laying off a road be Granted and that the said Petitioners do work on the said Intended road & that Nathaniel Hockett & John Bostick be Surveyors.

12 June 1746 O.S., p. 130
Roads & c.
Ordered the Surveyors of the Several High ways within this County be Continued According to their former Orders./.

10 July 1746 O.S., p. 139
Road.
Alexander M[t].Gomery and Others have leave to Clear a road According to theyr Petition Lodged in the Office, provided the said Petitioner do not carry their said Road through any fenced grounds.

14 Aug. 1746 O.S., p. 142
Road.
Ordered that a Road be Cleared from Beards Road on the Ridge between Appomattox and Willis's the nearest and best way to Albemarle Court House And alls that the Old road from Brooks Mill to John Bosticks be kept Open and that John Gannaway Stephen Sanders Joel Walker Daniel Low Edmond Gray and John Childers with their Male Tiths do Clear the same John Childers Overseer./.

14 Aug. 1746 O.S., p. 143
Road./.
Ordered a Road from Nicholas Davis's Plantation at the Blue Ridge falls to Beaver Creek be Opened and that the Male Tiths of Nicholas Davis William Stith John Bolling and George Stoball do clear the same. John Stone overseer It is likewise Ordered the said Road be Carried from thence to the South End of Slate River Mountains and that the Hands on the South Side the main river between Beaver Creek and Slate River Mountian do Clear the same James Christian Overseer and that the said Road be carried from thence a Cross Slate River near the Mouth of the Great Creek and that Allen Howard do mark off the said Road./.

16 August 1746 O.S., p. 164
Road & c.
Ordered Allen Howard James Daniels and Thomas Ballow Gent[n] or any two of them do View the Road from the Place William Battersby Road turns out of the New Road to the River, and likewise that they View the River to see if there is a Convenient place to Settle a Ferry, to the Court House and that they report the same to next Court & c

16 Aug. 1746 O.S., p. 164
Road.
Ordered the Road from Jones foard on Tye River be Continued to Swans Creek Mountain and from thence to the Mouth of Swans Creek and that the Secretarys Road be Continued from the Meadows to join John Harveys Road, and that the Widow Johnsons hands and the Inhabitants of Buffiloe and Tye River be Aded to the hands formerly Appointed for Clearing the same, and that William Cabell Gent. be Overseer thereof;--

16 August 1746 O.S., p. 164

Road

Charles Lynch and Edwin Hickman Gent[n]. are Ordered to View the Road to the Secretarys foard on the North River and that they Order the same to be turned as they think Proper.

16 Aug. 1746 O.S., p. 164

Road

Ordered the hands of Richard Randolph John Coles Joseph Fane and James Hill be Aded to the Hands formerly Appointed to work under Thomas Jobling.

16 Aug. 1746 O.S., p. 164

Road.

The Road from David Lewis Road to the late Secretarys Mill is Ordered to be kept in repair by the Secretarys hands that live above the Mountain the hands of Thomas Sowell and by the Inhabitants on Biskett Run Thomas Sowell Overseer. & c

16 Aug. 1746 O.S., p. 165

Road.

Ordered that a Road be cleared from the late Secretarys Mill to the Court House and that the Hands of Joshua Fry Castleton Harper John Burns Benjamin Whites James Lee and Thomas Butler do Clear the same. Castleton Harper Overseer.

16 Aug. 1746 O.S., p. 165

Road./.

David Lewis With the Hands formerly Appointed to work under him are Ordered to Continue their Road to the Secretarys Foard./.

16 Aug. 1746 O.S., p. 165

Road...

James Taylor with the hands formerly Appointed to work under him are Ordered to Clear a Road from his road to Martin Kings Road./.

12 Sept. 1746 O.S., p. 172

Road.

Ordered a Road be Cleared from Buffiloe Island to Harveys Road and the Second falls on Tye River by the Hands that are now on Harveys Road and that James Christian be Overseer of the same in the room of John Graves. & c

12 Sept. 1746 O.S., p. 173

Road....

Ordered the Road from Meridith Mannings to the Court House be kept open according to a former Order and that the hands of John Lee William Rickel James Goss Adrian Angle Hugh Green William Louhoon John Sharp Meridith Manning & Gideon Marr do Clear the same and that Gideon Marr be overseer & c

10 Oct. 1746 O.S., p. 200
Road & c
Ordered A Ferry be kept from the Land of William Battersby Gen[t] to the mouth of Totier Creek from on the Land of William Battersby Gent to the road from the Court House be turned by M[r]. Battersbys House and that John Ladd Noble Ladd and the Hands of William Battersby do Open the same William Battersby Overseer./. & c

13 Nov. 1746 O.S., p. 201
Road.
Ordered William Harris Gent be Overseer of the Road from Rock fish River to the Court House in the room of W[m]. Cabell Gent.

13 Nov. 1746 O.S., p. 201
Road.
Ordered Leonard Ballow be Overseer of the Ridge Road from the Place Cabells Road Interupts Freelands tract till they meet the Other Gang

13 Nov. 1746 O.S., p. 201
Road.
Ordered a Road from Burks Path on the North side the North River be Carried over Barringers Creek the nearest and best way to the Road against the Long Island Creek and that it be Cleared by the same Gang William Moseby Overseer.

13 Nov. 1746 O.S., p. 202
Road.
Ordered a Road be marked from Slate River to Glovers Road by Samuel Jordan Gent. in Stead of a Road formerly marked by Scruggs and that the said Road from the County Line at Phineas Glovers to Buckingham Path at William Webbs be Cleared by the Male Tithables of Isaac Bates James Daniel James Nivels and Richard Taylor Abraham Childers Overseer. And from the said Path to Slate River by the Male Tithables of William Cannon John Cannon Richard Cocke and All Other the Male Tiths between the mouth of Slate River and Isaac Bates that are not allready imploy'd on some other Road. and that Thomas Fouts be Overseer./.

12 Feb. 1746 O.S., p. 214
Road./.
The Viewers appointed to View the Road from the North Garden through Ivy Creek to Rock fish Road this day made their Report Ordered the male Tithables of Col°. Robert Lewis Thomas Walker William Bramham William Bustard and John Lyon do Clear the same John Lyon overseer

12 Feb. 1746 O.S., p. 214
Road./.
Ordered a Road be cleared from Holladay River to Red Oak House And that the Male Tithables of Phillip James Benjamin Mathews David Patteson John Jennins Thomas Lee Paul Chils and William Chambers do Clear the same William Chambers Overseer./.

12 Feb. 1746 O.S., p. 215
Road.
On the motion of John Biby leave is given him to Open a Road From the Mill to the Three forks of the Byrd./.

12 Feb. 1746 O.S., p. 215
Road./.
Ordered a Road be Cleared from the mouth of Tye River to the Tobacco Row Mountains and that all the Inhabitants between Buffiloe River Buffiloe Ridge and on the south River above the mouth of Harris's Creek do Clear the same. John Harvey Overseer

13 Feb. 1746 O.S., p. 219
Road./.
Ordered a Road be Cleared from Charles Caffreys on the Black water to Cross the Fluvanna to John Harveys Road, that Charles Caffrey John Hunt Thomas Hunt Martin Kelley Thomas Cooper John Caffrey & Charles Lynch's hands on the Fluvanna do Clear the same Chas. Lynch Overseer.

13 Feb. 1746 O.S., p. 219
Road.
Ordered Arthur M^c.Daniel be Overseer in the Room of William Cabell Gent. from the head of Sycomore Island Creek to Glovers Road and that the hands of William Nowland M^{rs}. Patteson John Ripley James Gates Thomas Turpin Abraham Smith Anthony Binnis John Gordon Sacheverel Whitebread Samuel Baily and Robertson Bailys hands do Clear the same

12 March 1746 O.S., p. 234
Road./.
Ordered the Old Road from John Bosticks to John Hodnetts be Cleared and from John Hodnetts along Dabbs's Path to the County Line and that the Male Tithables of William Gray John Sanders John Hodnett Daniel Rowand John Row do Clear the same John Sanders Overseer./

12 March 1746 O.S., p. 235
Road.
Fransis Baker is hereby Appointed Surveyor of the Mount Road from Number Twelve to the County Line. The former Gang. & c

13 March 1746 O.S., p. 245
Road./.
Ordered William Allen be Overseer of the Horn Quarter Road to the County Line and that the Male Tithables of Joseph Smith Robert King Reynee Shatteen William Howell Joseph Adcock William & Joseph Crie Robert Holt Anthony Sharroon Mathew Earps Patrick Obryan John Palmer John Hubard and William Sands do work Under the said William Allen in Clearing the Same./.

13 March 1746 O.S., p. 245
Road./.
Ordered a Road be Opened from the mouth of Totier to the Court House and that William Harris Gent: with the Gang formerly Ordered to work under him do clear the same

9 April 1747 O.S., p. 268
Road./.
Ordered Lazarus Dameron with his Three Sons Thomas Tindall Benjamin Tindall John Clark Charles Bond & his Tithables Richard Hall John Hall Thomas McDaniel and his hands do clear the Road from the Court House Shepards Ford Lazarus Dameron Overseer. & c

9 April 1747 O.S., p. 268
Road./.
Ordered the Male Tithables of Tye Chamberlain William Moor John Anthonys Henry Martin Thomas Crawley. Samuel Crawley Thomas Goolsby and John Goolsby do Clear the Road from the Court House to Martin Kings Road John Anthony Overseer. & c

14 May 1747 O.S., p. 271
Road. & c./.
The Petition for a Road from John W. Nortons (?) to the Court House is Granted. Ordered the said Petitioners do Clear the same. and that they be Exempt from Other Roads, Sanders M^t. Gomery Overseer. & c

14 May 1747 O.S., p. 271
Road.
The Inhabitants of Slate River have leave to Open a Road from Glovers Road along the ridge between Hatchers Creek and Little Buffiloe Creek to Alexander Trents on Willises Creek./.

9 July 1747 O.S., p. 293
Road./.
The Petition of William Diucuid and Others for Road to be Opened from the New Road from Slate River near M^r. Marrs's to the Court House is Granted the Male Tithables of William Diucuid Noble Ladd and John Ladd Are Ordered to Clear the same Noble Ladd Overseer. Noble Ladd & Gideon Marr are Ordered to mark out the same.

9 July 1747 O.S., p. 293
Road

Ordered Obediah Woodson be Surveyor of the Road Nathaniel Hockett was formerly Appointed Surveyor and that he have the same hands.

11 July 1747 O.S., p. 303
Road.

On the motion of John Biby leave is given him to Clear a Bridle Road from the said Bibys Mill to the Three forks of the Byrd. & c

14 Aug. 1747 O.S., p. 310
Road.

Ordered a Road be Cleared from Finly's Ferry to Samuel Stephens and that the Male Tithables of Margaret Finly James Freeland Robert Kile John Finly Jeremiah Whitney Thomas Thornhill John Coleman John Williams Richard Burks and Charles Burks do Clear the same. William Megginson Overseer. & c

12 Nov. 1747 O.S., p. 314
Road./.

Ordered that John Sorrell be Overseer of the Three Notched Road from N°. 12 in the Room of James Difir and that he have the same hands./.

12 Nov. 1747 O.S, p. 316
Road

Ordered that Samuel Shelton be Surveyor of the Road from Balingers Creek to Rock fish in the room of W^m. Harris Gent and that M^r. Stiths Tithables and all above Ballengers Creek do Assist the said Shelton in Clearing the same.

12 Nov. 1747 O.S., p. 316
Road.

Ordered John Goodwin be Surveyor of the Road from the Court House Ferry to the road he formerly Cleared in the room of Allen Howard Gent.

10 Dec. 1747 O.S., p. 329
Road.

William Fitzpatrick and Others have leave to Open a Road from Stephen Heards to the late Secretarys Road near John Burns Plantation.

11 Dec. 1747 O.S., p. 342
Road./. --

Ordered the Male Tithables belonging to the Late Secretary at Clear Mount do work on the Road from James Taylors Foard to Martin Kings Foard and that James Taylor be overseer/--

10 March 1747 O.S., p. 344
Road
Ordered a Road be cleared from the Green mountain Road near the head of Hogg Creek into the Court House Road below Mr Stiths Quarter And that the Hands of Col^o: Richard Randolph M^r. Stiths and William Harris's hands do Clear the Same W^m. Harris Overseer./.

10 March 1747 O.S., p. 344
Road./.
Reny. Shatteen is Appointed Surveyor of the Horn Quarter Road from the County line to Glovers Road in the room of William Allen and the Tithables of John Parmer Joseph Adcock Robert King Joseph Smith John Carnys William Carnys Aaron Carver Thomas Gresham Robert Holt Archibald Cary and David Bell do work under the said Shatteen on the said Road./.

10 March 1747 O.S., p. 345
Road./.
Matthews Ayres Patrick Obryan and William Olle are Ordered to be Added to the Gang from Glovers Road to Slate River W^m. Allen Overseer.

10 March 1747 O.S., p. 345
Road
W^m. Mills is Appointed Surveyor of the Road in the room of Robert Davis.

10 March 1747 O.S., p. 345
Road
Ordered a Road be Opened from Holladay River to meet Otter River and that the same Surveyor and hands do Continue the same.

10 March 1747 O.S., p. 345
Road.
Ordered Wllllam Hamner be Surveyor of the road from the Late Secretary's Mill to the Court House in the room of Castleton Harper & that he have the same hands

10 March 1747 O.S., p. 345
Road.
the Tithables of Henry Hamilton and Samuel Moor are Ordered to be Added to Gideon Morris Gant to work on his Road & c

10 March 1747 O.S., p. 347
Road.
John Lewis is Appointed Surveyor of the Road from Court House to Ballingers Creek in the room of William Harris. Except M^r. Nichols's Mr Harris'. Ordered M^r. Nichols's hands be Added to M^r. Shelton's Gang. & c./.

11 March 1747 O.S., p. 347

Road.

Ordered Samuel Spencer be Surveyor of the Road From the Fluvanna to Freeland's tract in the room of William Cabell./. And that Philip Davis be Overseer of the Road from Jones's Foard on Tye River to Swans Creek Ferry./. in the room of William Cabell and that Philip Morris be Overseer of the Road from Rock Fish river to Havis Road in the Room of W^m. Cabell and that the Tithables of Col^o. Lomax Harmer & King & the Settlements on Cabell's Land on Buffiloe & Tye River do Clear the Same.

11 March 1747 O.S., p. 348

Road

Ordered the Road from Freeland's tract be Continued by John Goodwin's and from thence to Marrs' Road be Slate River and that Arthur M^c.Daniel & the hands Appointed to work Under him do Clear the Same Ordered few Moor's Tiths work on the road Under gideon Marr./.

12 March 1747 O.S., p. 358

Ordinary Licence & ferry Bond./.

On the motion of William Cabell leave is given him to keep an Ordinary at his Ferry in this County on his giving Bond & c who together with Benjamin Harris his Security Enter into and Acknowledged their Bond & c for the said Cabell's keeping an Ordinary according to Law and the said Cabell also acknowledges a Ferry Bond & c

12 March 1747 O.S., p. 359

Road

Ordered a Road be Cleared from the Court House Road below Ballenger's Creek the Best way to Martin King's foard on the North River and that John Anthony Gent: be Overseer.

12 March 1747 O.S., p. 359

Road.

Ordered the Hands at Turpin's Quarter be added to Gideon Marrs' Road./.

12 May 1748 O.S., p. 360

Road.

Ordered John Ried David Lewis and John Wood or any two of them do View the way from the great Mountains to Morrison's Road & make report & c

12 May 1748 O.S., p. 361

Road.

Ordered David Lewis and his Gang do Clear a road from Charles Lynch's : ferry road the nearest and most Convenient way to the Court House./.

13 May 1748 O.S., p. 369
Road.
Ordered Joshua Fry & Charles Lynch Gent. do Apply to Louisa Court that the Road from King's foard on the Rivanna may be Continued from the County Line to Louisa Court House.

9 June 1748 O.S., p. 372
Road./.
On the Motion of Giles Aligree It is Ordered the Road Over MeChunk Creek by his House be turned the Old way. & c

9 June 1748 O.S., p. 372
Road./.
William Cabell James Christian and John Harvie or any two of them are Ordered to View the Road from the mouth of Tye River to the Lower foard on Rock fish & make their report to next Court

9 June 1748 O.S., p. 372
Road.
The dispute between the Inhabitants of Rock fish about a Road from Rock fish Gap to Morrison's Road is Continued for further argument til Septe^r. Court./. next.

11 Aug. 1748 O.S., p. 399
Road.
Joel Walker is Appointed Overseer of the Road from Chiles's to the Plantation where John Bostick formerly lived the same hands & c

11 Aug. 1748 O.S., p. 399
Road
A former Order for a Road made August Court 1746 John Childers Over seer is Ordered to be discontinued the same being found to be Inconvenient

11 August 1748 O.S., p. 399
Road
Ordered the Sheriff in the Several Surveyors of High Ways within this County Notice to Attend this Court in September next and that the Clerk in the mean time make out Lists of the Several Surveyors with in this County

August 13, 1748 O.S., p. 414
Road
On the motion of Richard Burks leave is given him to Clear a Road from John Coleman's to John Beard's Road at the head of appomattox

INDEX -ALBEMARLE COUNTY ROAD ORDERS

Note: This index is arranged by subject: Personal Names; Ferries; Fords; County Government; Houses; Mills; Mountains, Gaps, Passes, Valleys, etc; Rivers, Creeks, Runs, Islands, Falls, etc.; Roads, Paths, etc.; Quarters; Signposts, Marked Trees, etc.; Road Surveyors Gangs; Tithables Listed by Owner

Personal Names:

Robert Adams, 13
Mr. Adam's, 12
David Addams, 5
Joseph Adcock, 18, 20
Giles Aligree, 22
William Allen, 8, 10, 13, 18$^{(2)}$, 20$^{(2)}$
Charles Anderson, 13
Robert Anderson, 6
Adrian Angle, 15
John Anthony, 7$^{(2)}$, 11$^{(2)}$, 18$^{(2)}$, 21
Matthew Ayres, 20
Robertson Bailys, 17
Samuel Baily, 17
Fransis Baker, 17
Leonard Ballow, 16
Thomas Ballew (Ballow, Bellew), 6, 7, 8, 14
Capt. Bellew's, 7
William Barnett, 13
Isaac Bates, 16$^{(2)}$
William Battersby, 16$^{(4)}$
John Beard's, 22
David Bell, 20
Thomas Bibb, 12$^{(2)}$
John Biby, 17, 19
Anthony Binnis (Bennins?), 17
John Bolling, 14
Charles Bond, 18
John Bostick, 13, 14, 17
William Bramham, 16
Charles Burks, 19
Richard Burks, 19, 22
Samuel Burks, 7, 12
Henry Burns, 12
John Burns, 15, 19
William Bustard, 16
Thomas Butler, 15
William Cabell, 6$^{(2)}$, 9, 14, 16, 17, 21$^{(5)}$, 22

Charles Caffrey, 17[2]
John Caffrey, 17
John Cambhill, 5
John Cannon, 7, 12, 16
William Cannon, 16, 20
William Carnys, 20
Aaron Carver, 20
Archibald Cary, 20
Miles Cary's, 12
Howard Cashe's, 9
Tye Chamberlain, 18
William Chambers, 17[2]
Abraham Childers, 16
John Childers, 14[2], 22
Paul Chils, 17
James Christian, 15, 22
John Clark, 18
Richard Cocke, 16
John Coleman, 19, 22
John Coles, 15
Thomas Cooper, 17
John Cowan, 6
Samuel Crawley, 7, 18
Thomas Crawley, 18
Joseph Crie, 18
William Crie, 18
Lazaros Dameron, 9, 11, 13, 18[2]
Lazarus Dameron's Three Sons, 18
James Daniels, 14, 16
Robert Daulton, 8
Nicholas Davis, 14[2]
Philip Davis, 21
Robert Davis, 20
James Defoe (Difir), 5, 19
John Dickey, 6
Anthony Dinnies, 13
William Diucuid, 18[3]
Hugh Dobbins, 11
Mathew Earps, 18
Jno: Edmonds, 7
Joseph Fane, 15
James Fenly, 8[2]
John Fenley, 8[2]
John Finly, 19
Margaret Finly, 19
William Fitzpatrick, 19

Thomas Fouts, 16
James Freeland, 19
Joshua Fry, 15, 22
John Gannaway, 14
James Gates, 17
Phineas Glovers, 16
Samuel Glover, 12[2], 13
John Goodwin, 19, 21
John Goolsby, 18
Thomas Goolsby, 18
John Gordon, 17
James Goss, 15
John Graves, 10, 15
Mathew Graves, 10
Edmond Gray, 9, 14
William Gray, 17
Hugh Green, 15
Thomas Gresham, 20
John Hall, 18
Richard Hall, 18
Harry Hamilton, 20
William Hamner, 20
Colo. Lomax Harmer, 21
Castleton Harper, 15[2], 20
Benjamin Harris, 21
Mr. Jno. Harris, 7
William Harris, 11, 16, 18, 19, 20[3]
Mr. Harris', 20
John Harvey (Harvie), 6, 14, 17[2], 22
Mr. Harreys, 11
John Heard, 12
Stephen Heard, 19
John Henderson, 5
Majr. John Henry's, 12
Thomas Herbert, 13
Edwin Hickman, 15
Moses Higginbottom, 11
James Hill, 15
George Hilton, 9, 10
William Hind, 8
Nathaniel Hockett, 13, 19
John Hodnett, 17[3]
Daniel Holladay, 8
Robert Holt, 18, 20
Doctr. Hopkins's, 10
Old Dr. Hopkin's, 12

Allen Howard, 6, 7[2], 14[2], 19
William Howell, 18
John Hubard, 18
Ellis Hues', 7
Thomas Hughs, 5
John Hunt, 17
Thomas Hunt, 17
John Hunter, 13
John Isham, 7
Phillip James, 17
Peter Jefferson, 6
John Jennins, 17
John Johns, 8
Widow Johnsons, 14
Thomas Jones, 8
Thomas Joplin (Jobling), 8, 15
Samuel Jordan, 16
Martin Kelley, 17
Jno Kerrs, 11
John Key, 8[3], 11
Robert Kile, 19
Martin King, 7[3], 12, 15, 18, 19, 21
Robert King, 13, 18, 20
Joseph Kinkead, 10
John Ladd, 16, 18
Noble Ladd, 16, 18[3]
Daniel Lattimore, 5
Charles Lavinder, 7, 10
John Lawson, 6
James Lee, 15
John Lee, 15
Thomas Lee, 17
David Lewis, 6, 15[2], 21[2]
John Lewis, 13, 20
Colo. Robert Lewis, 16
Capt. Lewis, 11
William Little, 6
William Louhoon, 15
Daniel Low, 14
Charles Lynch 5, 6, 7, 9[2], 10, 15, 17[3], 22
John Lyon, 11, 16[2]
Robert Mannely, 6
Meriday (Meredith) Mannings, 10, 15[2]
Gideon Marr (Marris), 15[2], 18, 20, 21
Jeremiah Marris, 6
Henry Martin, 12, 18

Capt. James Martin, 12
Thomas Martin, 13
Colo. Martin's, 12
Benjamin Mathews, 17
John Mc.Cord, 5
Arthur Mc.Daniel, 17, 21
Thomas McDaniel, 18
Micah Mc:Dowell, 6
Mirlock Mc.Dowell, 6
Andrew Mc:Williams, 13
William Megginson, 19
Capt. Thomas Meriwether, 7
Colo. Wm. Meriwether, 11
Mr. Meriwether, 11
Wm. Mills, 20
Samuel Moor, 20
William Moor, 18
Philip Morris, 21
Thomas Morrison's, 5
William Morrison, 5, 12
William Moseby, 16
Alexander Mt.Gomery, 14
Sanders Mt.Gomery, 18
Mr. Nichols's, 20[2]
James Nivels, 16
Capt. Nevell's, 8
John W. Nortons, 18
William Nowland, 17
Patrick Obryan, 18, 20
William Olle, 20
Anthony Osbrook, 6
John Palmer, 18
John Parmer, 20
David Patteson, 17
Mrs. Patterson, 17
Richard Randolph, 15
Colo. Richard Randolph, 20
William Rickel, 15
John Ried, 21
John Ripley, 17
Robert Rose, 6
Daniel Row, 17
John Row, 17
William Sands, 18
John Sanders, 17[2]
Stephen Sanders, 14

Daniel Scott, 8, 9[3]
Scruggs (overseer), 16
John Sharp, 15
Anthony Sharroon, 18
Reynee Shatteen, 18, 20
William Shaw, 6
Samuel Shelton, 19
Abraham Smith, 17
Joseph Smith, 13, 18, 20
Larkin Smith, 8
John Sorrell, 19
Thomas Sowell, 15[2]
Samuel Spencer, 21
Samuel Stephens, 19
William Stith, 14
the Revd: Mr. Stith, 7
Mr. Stiths, 19, 20[2]
George Stoball, 14
Davis Stockton, 10
John Stone, 14
William Sudworth, 11
George Tayler, 12
James Taylor (Tayler), 6, 15, 19[2]
Richard Taylor (Tayler), 8, 16
Joseph Thomson, 10
Mr. Thomsons, 10
Thomas Thornhill, 19
John Thrasher, 6
Benjamin Tindall, 18
Thomas Tindall, 18
Alexander Trents, 18
Thomas Turpin, 17
Joel Walker, 14, 22
Thomas Walker, 11, 12, 16
Andrew Wallace, 6
William Wallace, 6
William Webbs, 16
Benjamin Wheeler, 5[2]
Sachererel Whitebread, 17
Benjamin Whites, 15
Jeremiah Whitney, 19
John Williams, 19
William Womack, 8
Andrew Wood, 10
Archibald Woods, 6
John Wood, 21

Micah Woods junr., 6
William Woods, 10, 13
Benjamin Woodson, 9, 10[2], 12
Obediah Woodson, 19
John Woody, 5

Ferries:
Cabell's Ferry, 21
Cabell's Ferry Bond, 21
Court House ferry, 14, 19,
Ferry, 8
Finly's Ferry, 19
Charles Lynch's Ferry, 9, 11
Lynchs Ferry Bond, 9
Scotts Ferry, 8
Swans Creek Ferry, 21
Totier Creek Ferry, 16

Fords:
Robert Davis's ffoard, 9
Jones foard, 14, 21
Capt. Charles Lynches foard, 6
Meriday Mannings foard, 8
Martin Kings Foard, 19, 21, 22
ffoard on Mitchams River, 9, 10
Lower foard on Rock Fish River, 22
Rock fish foard, 8
Rooks's Foard, 12[2]
Secretarys ffoard, 8, 10, 11, 15
Shepards ffoard, 9, 10, 11, 13, 18
James Taylors Foard, 19
ffoard on Tye River, 6
Benjamin Woodsons ffoard, 9, 10

County Government:
Clerk, 6
County Line, 5, 7, 12, 13, 16, 17[2], 18, 20, 22
Court House, 6[2], 7[4], 8[2], 9[2], 10, 11, 13, 14[2], 15[2], 16[2], 18[5], 19, 20[3], 21[2]
Court House Landing, 8
Louisa Court House, 7, 22
the Office, 14
Ordinary Licences
        Cabell's, 21
        Scott's, 9
Petitions, 5, 6, 9[2], 10, 11, 12, 13, 14, 18
Sheriff, 22

Lists of Surveyors of Roads, 22

Houses:
Giles Aligre's House, 22
Mr. Battersbys House, 16
Joseph Kinkead's House, 10
Red Oak House, 17
Joseph Thomson's Old House, 10

Mills:
the Mill (Bibys), 17, 19
Brooks Mill, 14
Moses Higginbottoms Mill, 11
John Keys Mill, 8, 11
Secretarys Mill, 15[2], 20
Davis Stockton Mill, 10

Mountains, Gaps, Passes, Valleys, etc.:
Ridge between Appomattox and Willis's, 14
Blue Mountains, 9, 10
Blue Ridge, 14
Buffiloe Ridge, 17
ridge between Hatchers Creek, and Little Buffiloe Creek, 18
Ivey Creek Pass, 11
the Meadows, 14
the Mountains, 12, 15
great Mountains, 21
north Garden, 11, 16
Red Mountains, 9
the Ridge, 13
Rock fish Gapp, 9, 22
Secretarys Mountains, 11
Slate River Mountains, 14[5]
Swans Creek Mountain, 14
Tobacco Row Mountains, 17
Michael Wood's Gapp, 5, 9, 10

Plantations:
John Bostick's Plantation where he formerly lived, 22
John Burns Plantation, 19
Cabells Land, 21
Chiless, 22
Clear Mount, 19
John Coleman's, 22
Nicholas Davis's Plantation, 14
Freelands Tract, 16, 21[2]

30

John Goodwin, 21
John Henderson's Land, 5
William Harris's Plantation, 11
Stephen Heards, 19
Leak's Plantation 6
Mr. Marr's, 18
Robert Rose's Plantations, 6
Samuel Stephens, 19
Alexander Trents, 18

Rivers, Creeks, Runs, Islands, Falls, etc.:
Appomattox River, 14, 22
Ballengers Creek, 19[2], 20, 21
Barrengers Creek, 11, 16
Beaver Creek 14[2]
Biskett Run 15
Blackwater River, 17
Blue Ridge falls, 14
Buck Island Creek, 12
Little Buffiloe Creek, 18
Buffiloe Island, 15
Buffiloe River, 10, 11[2], 14, 17, 21
Three forks of the Byrd (creek), 17, 19
Cunningham Creek, 10
Fluvanna River (James), 7[3], 10, 21
Great Creek, 14
Green Creek, 11
Hardwar River, 5, 7[2], 8, 10[3], 11
Harris's Creek, 6[2], 10[2], 17
Hatchers Creek, 13, 18
Hogg Creek, 20
Holladay River, 17, 20
Ivy Creek, 11, 16
James River, 6, 7
ffork of James River, 9
Long Island Creek, 16
Me Chunk Creek, 22
Mitchams River, 6, 9[3], 10[2]
North River (Rivanna), 9, 10, 15, 16
Otter River, 20
Point of fork, 10
Rivanna River, 22
Rock fish River, 8[2], 16, 19, 21, 11
Slate River 7[2], 8, 12[2], 13, 14[4],16[3], 18, 20, 21
South River (James) 11, 17
Sycomer Island Creek, 7, 17

Swans Creek, 8, 14, 21
Totier Creek, 16, 18
Tye River, 6[(4)], 10[(2)], 11, 14, 15, 17, 21[(2)], 22
Second falls on Tye River, 15
Willises Creek, 18
Willis's River, 14

Roads, Paths, etc.:

John Anthony's Road, 12[(2)]

Road from Balingers Creek to Rock fish, 19

William Battersby Road, 14

Road from the Place William Battersby Road turns out of the New Road to the River, 14

John Beards' Road, 22

Beards Road on the Ridge between Appomattox and Willis's, 14

Road from Beards Road on the Ridge between Appomattox and Willis's to Albemarle Court House, 14

Road from Beaver Creek to the South End of Slate River Mountains, 14

Capt. Bellew's Road, 7

High Way From the place Capt. Bellew's Road comes into Slate River to Glovers Road, 7

Road from Bibys Mill to the Three forks of the Byrd, 17, 19

Old Road from John Bosticks to John Hodnetts and along Dabbs's Path to the County Line, 17

Old road from Brooks Mill to John Bosticks, 14

Buckingham Path at William Webbs, 16

road from Buckingham Path to Slate River, 16

Road from Buffiloe Island to Harveys Road and the Second falls of Tye River, 15

Burks Path, 16

Road from Burks Path on the North side the North River over Barringers Creek to the Road against Long Island Creek, 16

Cabells Road, 16

Ridge road from the Place Cabells Road Interupts freelands tract, 16

Road from Charles Caffreys on the Blackwater to Cross the Fluvanna to John Harveys Road, 17

Road from Chiles's to the Plantation where John Bostick formerly lived, 22

Road from John Coleman's to John Beard's Road at the head of Appomattox, 22

Road from the County line at Phineas Glovers to Buckingham Path at William Webbs, 16

road from the Court House, 16

Road from the Court House to Ballingers Creek, 20

High Way from the Court House to the Road formerly Cleared by Allen Howard Gent: and from thence to be carried by the said Bellew to Slate River, 7

High Way on the South side of River from the Court House up the Country, to the Road formerly Cleared by the said Howard and thence to Sycomer Island Creek, 7

High Way from the Court House to Martin Kings Road, 7, 18

Road from the Court House to Meriday Mannings foard on Slate River, 8

Road from the Court House to Shepards foard on Hardwar River, 11, 18

Road from the Court House to Shepards foard to be Carried round the Ridge, 13

Road from the Court House to Shepards ffoard from thence to Doctr. Hopkins's Road, 9, 10

Road from the Court House Ferry to the road John Goodwin formerly Cleared, 19

Court House Road, 20, 21

Road from the Court House Road below Ballenger's Creek to Martin King's foard, 21

Dabbs's Path, 17

Road from Nicholas Davis's Plantation at the Blue Ridge falls to Beaver Creek, 14

Road from Robert Davis's ffoard to Howard Cashe's, 9

Highway from D.S. to Mitchams River, 6

High Way from D.S. to W.M., 5

D.S. Road, 9, 11

Road from D.S. Road to Capt. Charles Lynch's ferry, 11

Road from Finly's Ferry to Samuel Stephens, 19

Road from the Fluvanna to Freeland's tract, 21

Road from Freeland's tract by John Goodwin's and to Marr's Road, 21

Road from Samuel Glovers to Horn Quarter, 12

Road from Glovers Quarter to Slate River, 12

Glovers Road, 7, 13, 16, 17, 18, 20[(2)]

Road from Glovers Road along the ridge between Hatchers Creek and Little Buffiloe Creek to Alexander Trents on Willises Creek, 18

road John Goodwin formerly Cleared, 19

Green mountain Road, 20

Road from the Green mountain Road near the head of Hogg Creek into the Court House Road below Mr. Stiths Quarter, 20

Harris Road, 21

Road from William Harris's Plantation on Green Creek to the South River on the lower side of Barrengers Creek, 11

John Harveys Road, 11, 14, 15[(2)], 17

Road from Stephen Heards to the late Secretarys Road near John Burns Plantation, 19

Road from Moses Higginbottoms Mill a Cross Buffiloe to Mr. Harveys Road, 11

Road from Holladay River to meet Otter River, 20

Road from Holladay River to Red Oak House, 17

Doctr. Hopkins's Road, 9, 10[(2)], 12

Road from the Place it comes into Hopkins' road to the County line, 12

Horn Quarter Road, 13

Horn Quarter Road to the County line, 18

Horn Quarter Road from the County line to Glovers Road, 20

Road formerly Cleared by Allen Howard 7[(2)]

Road from Jones foard on Tye River to Swans Creek Mountain and on to the Mouth of Swans Creek, 14

Road from Jones's Foard on Tye River to Swans Creek Ferry, 21

High Way, from the End of A Road which is Cleared, to John Keys Mill, the nearest. and best way to the Late Secretarys ffoard, 8

Road Which is Cleared, to John Keys Mill, the nearest. and best way to the Late Secretarys ffoard, 8, 11

John Keys Road, 11

Road from King's foard on the Rivanna from the County Line to Louisa Court House 22

Martin Kings Road, 7[(3)], 12, 15, 18

Road from Joseph Kinkeads House to Davis Stockton Mill, 10

David Lewis Road, 15

David Lewis Road to go to Secretarys foard, 15

Road from David Lewis Road to the late Secretarys Mill, 15

Road against the Long Island Creek, 16

Charles Lynch's ferry road, 5, 21

road from Charles Lynch's ferry road to the Court House, 21

road over Capt. Charles Lynches foard, 6

Road from Meridith Mannings to the Court House, 15

Road from Meriday Mannings to Horn Quarter, 10

Road from Meriday Mannings foard on Slate River to Horn Quarter, 8

Gideon Morris (Marr's) Road, 20, 21[2]

Mc.Cords Road, 5

Highway from the uper end of Mc.Cords Road to Thomas Morrisons, 5

Road Over Me Chunk Creek by Giles Aligree's House to be turned the Old way, 9

Road from the ffoard on Mitchams River to Michael Woods' Gapp, 9, 10

Morrison's Road, 21, 22

Way from the great Mountains to Morrison's Road, 21

Mount Road from Number Twelve to the County Line, 17

Road from the north Garden through Ivey Creek Pass to Rock fish Road, 11, 16

Road from John W. Nortons to the Court House, 18

New Road to the River, 14

Road from Robert Rose's Plantations on Tye River to Leak's Plantation, 6

High Way from Rock fish to Hardwar, 8

(road) from the fork of Road to Rock fish foard, 8

Road from Rock fish to Morrison's Road, 22

Road from Rock fish Gapp to D.S. road, 9

Road from Rock fish River to the Court House, 16

Road from Rock fish river to Harris Road, 21

Rock fish Road, 11, 16

Road formerly marked by Scruggs, 16

Road to the Secretarys foard on North River, 15

High Way from the late Secretary's foard to Number 12, 5

Road from the Secretary's Foard to Twelve Mile tree, 10

Road from the late Secretarys Mill to the Court House 15, 20

Secretarys Road, 14, 19

Secretarys Road from the Meadows to John Harveys Road, 14

Road from Slate River to Glovers Road, 16

New Road from Slate River, 18

Road from the New Road from Slate River near Mr. Marr's to the Court House, 18

Road from the South End of Slate River Mountains a Cross Slate River near the mouth of Great Creek, 14

Road to Swans Creek, 8

Road from the head of Sycomore Island Creek to Glovers Road, 17

Road from James Taylors Foard to Martin Kings Foard, 19

Road from James Taylor's Road to Martin Kings Road, 15

Road from Joseph Thomsons Old House to Doctr. Hopkin's Road, 10

Three Notch'd Road, $5^{(2)}$, 7, $12^{(2)}$, 19

Three Notch'd Road from No. 12, 19

highway from Number 12: to Number 18, 5

Highway from Number 18 to the County Line on the Three Notched Road, 5

Road from Three Notch'd Road to Rooks's Foard, 12

road through the Land of John Henderson, from the Three Notch'd Road and to Hard way River, 5

Road from the place Martin Kings Road come into the Three Notched Road by Ellis Hues' to the County Line towards Louisa Court House, 7

four Chopt Road, 5

Road from the mouth of Totier to the Court House, 18

High Way from the ffoard of Tye River to the Court House, 6

road from the Mouth of Tye River to the Branches of Harris's Creek, 6, 10

Road from the mouth of Tye River to the Lower foard on Rock fish, 22

Road from the mouth of Tye River to the Tobacco Row Mountains, 17

Road from Benjamin Wheeler's into the four Chopt Road; to Wood's Gapp, 5

Quarters:
Glovers Quarter, 12
Horn Quarter, 8,10,12
Mr. Stiths Quarter, 20
Turpin's Quarter 21

Signposts, Marked Trees, etc.:
Signposts, 8
Height of Signposts, 8
D.S. Tree, 5, 6
W.M.(?), 5
Twelve Mile tree, 10
Number 12, 5[2], 17, 19
Number 18, 5[2]

Road Surveyors Gangs:
Jno: Anthony's Gang, 7
Jno Graves's Gang, 11[2]
David Lewis's Gang, 21
Gideon Morris (Marr's) Gang, 20
Mr. Shelton's Gang, 20

Tithables Listed by Owner:
Samuel Burks's Tithables, 7
uper part of Miles Cary's Hands, 12
Mr. Cary's hands at Hatchers Creek, 13
Samuel Crawley's Tithables, 7
Jno: Edmond's Tithables, 7
Mr: Jno. Harris' Tithables, 7
John Isham's Tithables, 7
King's Tithables, 21
Charles Lavinders' Tithables, 7
Charles Lynch's hands on the Fluvanna, 17
Capt. Thomas Meriwether's Tithables, 7

Moor's Tiths, 21
Capt. Nevell's Hands, 8
Mr. Nichols' hands, 20
Late Secretarys Tithables at Clear Mount, 19
Secretarys hands above the Mountain, 15
late Secretary's hands below the Mountains, 12
the Revd: Mr. Stith's Tithables, 7, 19
Mr. Thomsons Tlths, 10